Fossils And History:
Paleontology for Kids
(First Grade Science Workbook Series)

Speedy Publishing LLC
40 E. Main St. #1156
Newark, DE 19711
www.speedypublishing.com

Copyright 2015

All Rights reserved. No part of this book may be reproduced or used in any way or form or by any means whether electronic or mechanical, this means that you cannot record or photocopy any material ideas or tips that are provided in this book

Paleontology is the branch of biology that studies the forms of life that existed in former geologic periods, primarily by studying fossils.

The only direct way we have of learning about dinosaurs is by studying fossils.

The word fossil comes from the Latin word fossilis, which means, "dug up".

Fossils have been found on every continent on Earth.

Fossils are the remains of ancient animals and plants, the traces or impressions of living things from past geologic ages, or the traces of their activities.

Over long periods of time, these small pieces of debris are compressed (squeezed) and are buried under more and more layers of sediment that piles up on top of it.

Fossils of imprints may form, like casts of dinosaur footprints.

The fossil of a bone doesn't have any bone in it!

FOSSIL FUEL POWER STATION

Fossil fuels are essentially the remains of plants of animals. They provide us with a source of nonrenewable energy.

A fossilized object has the same shape as the original object, but is chemically more like a rock.

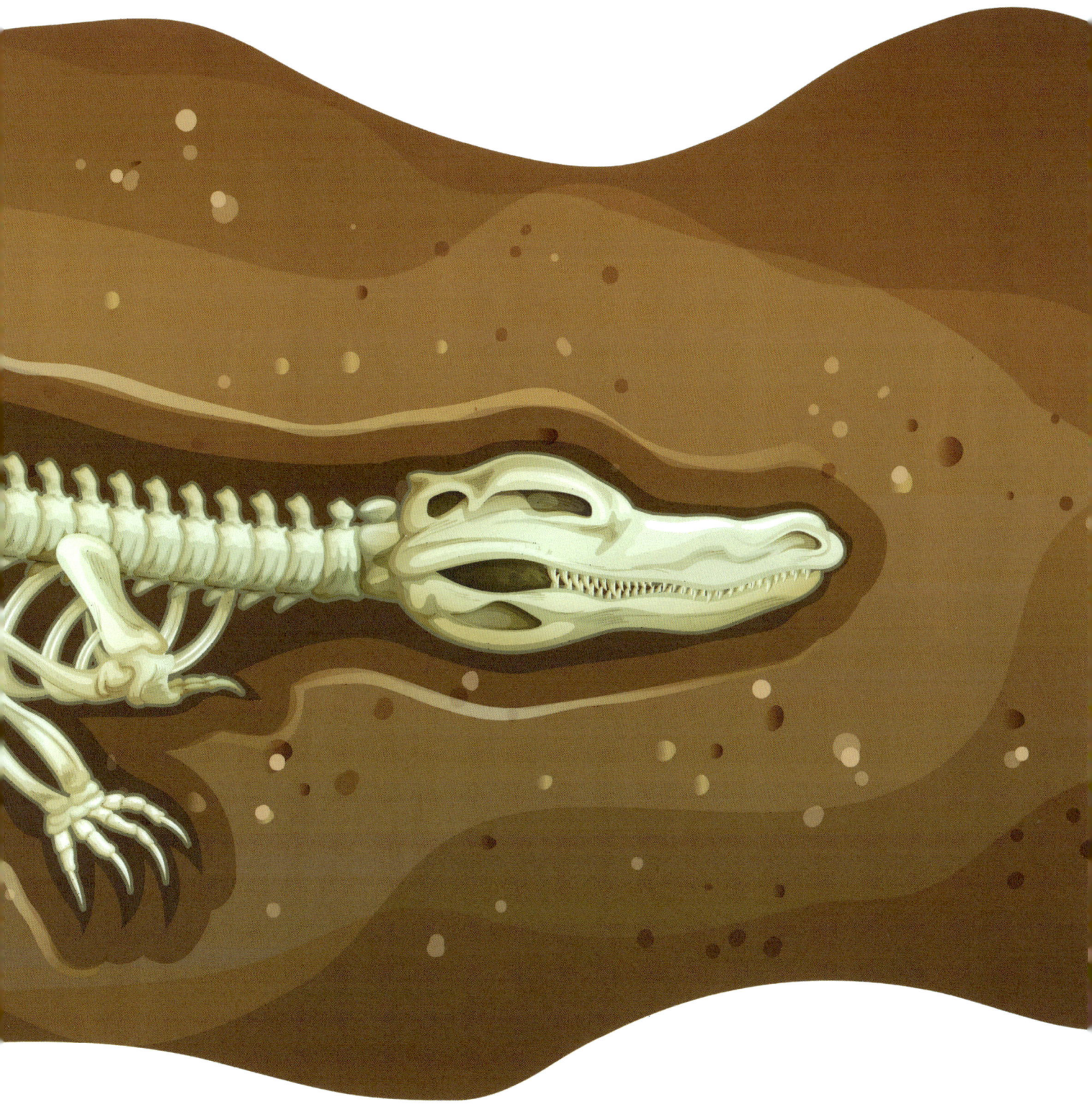

CONNECT THE DOTS & COLOR!

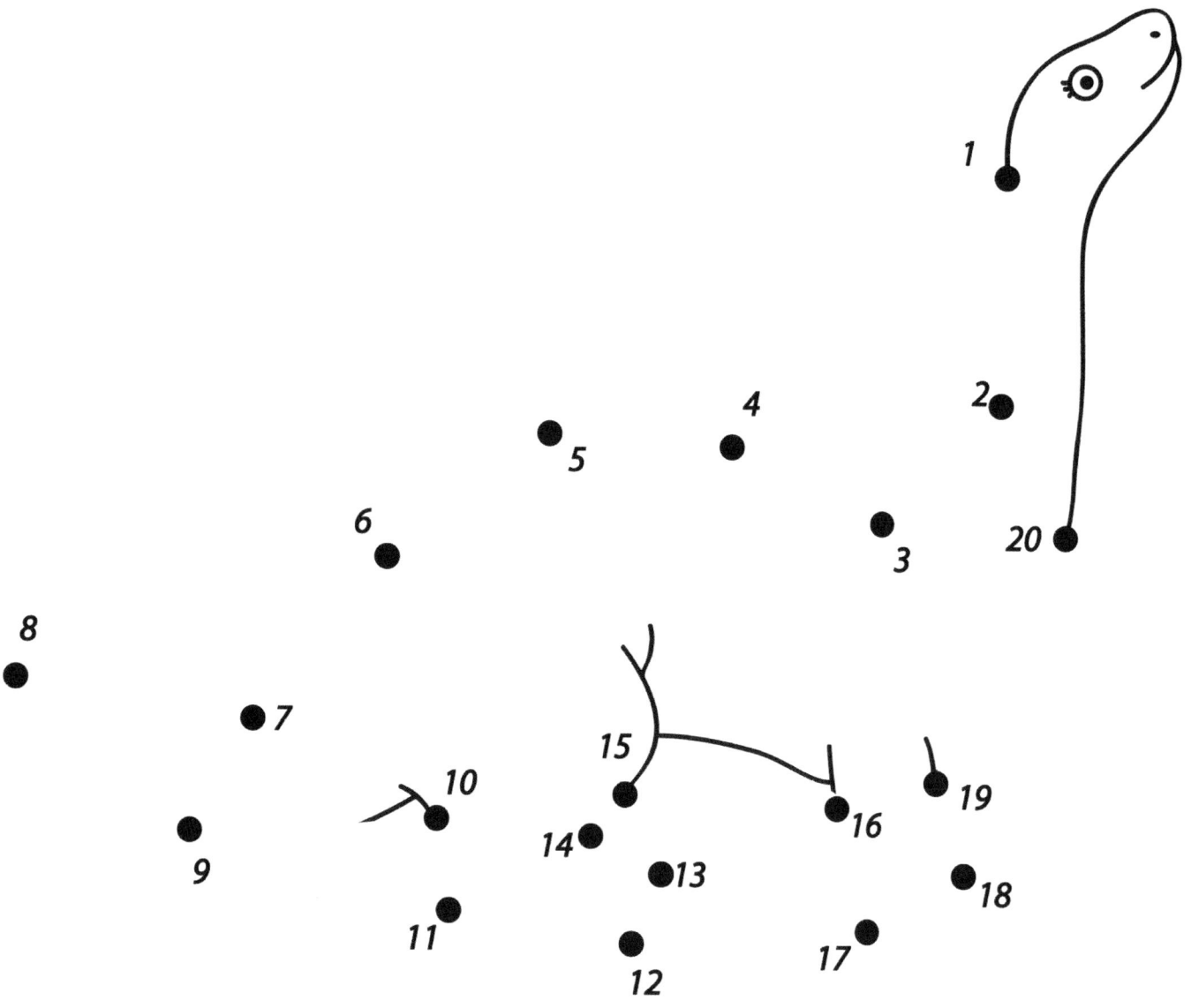

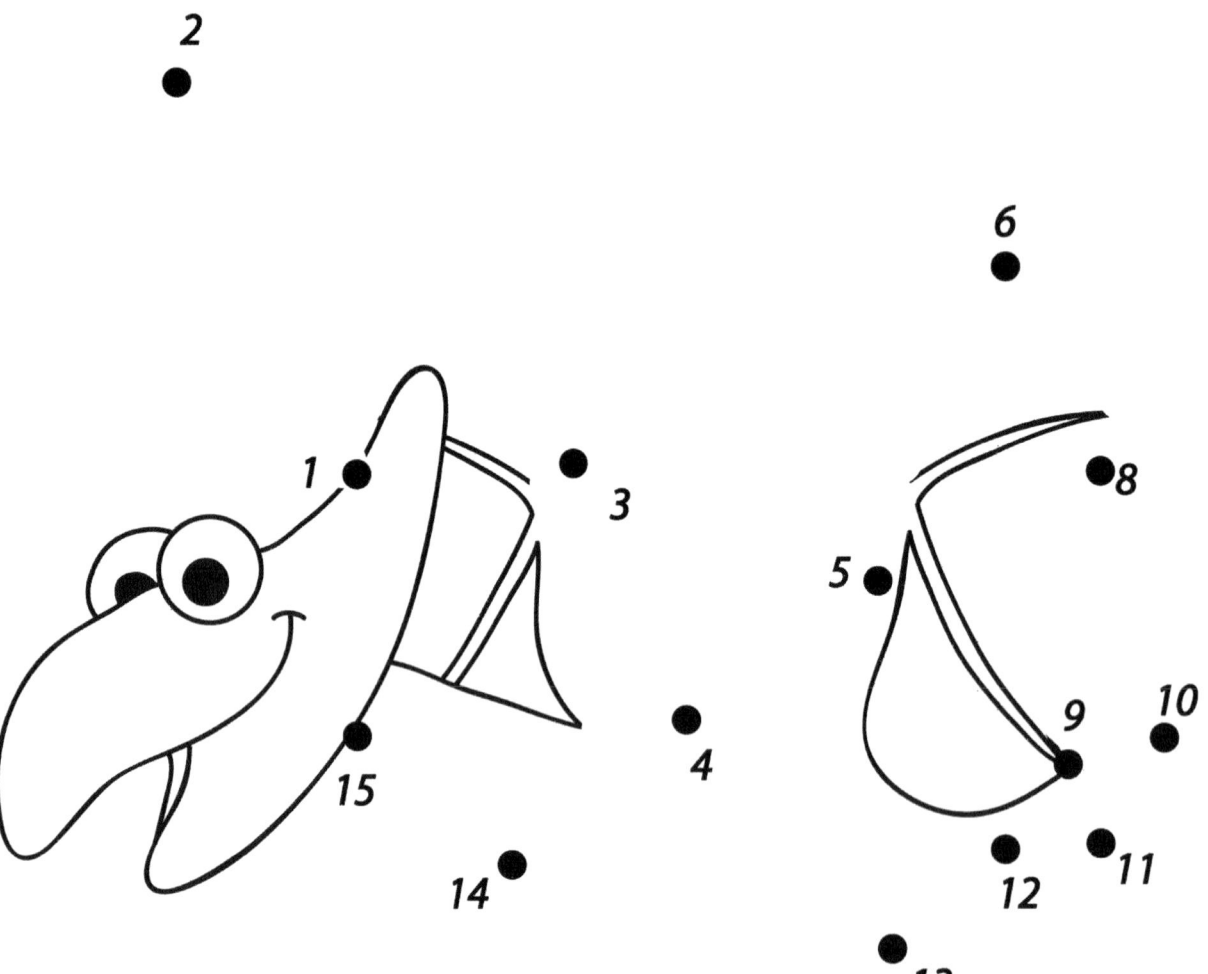

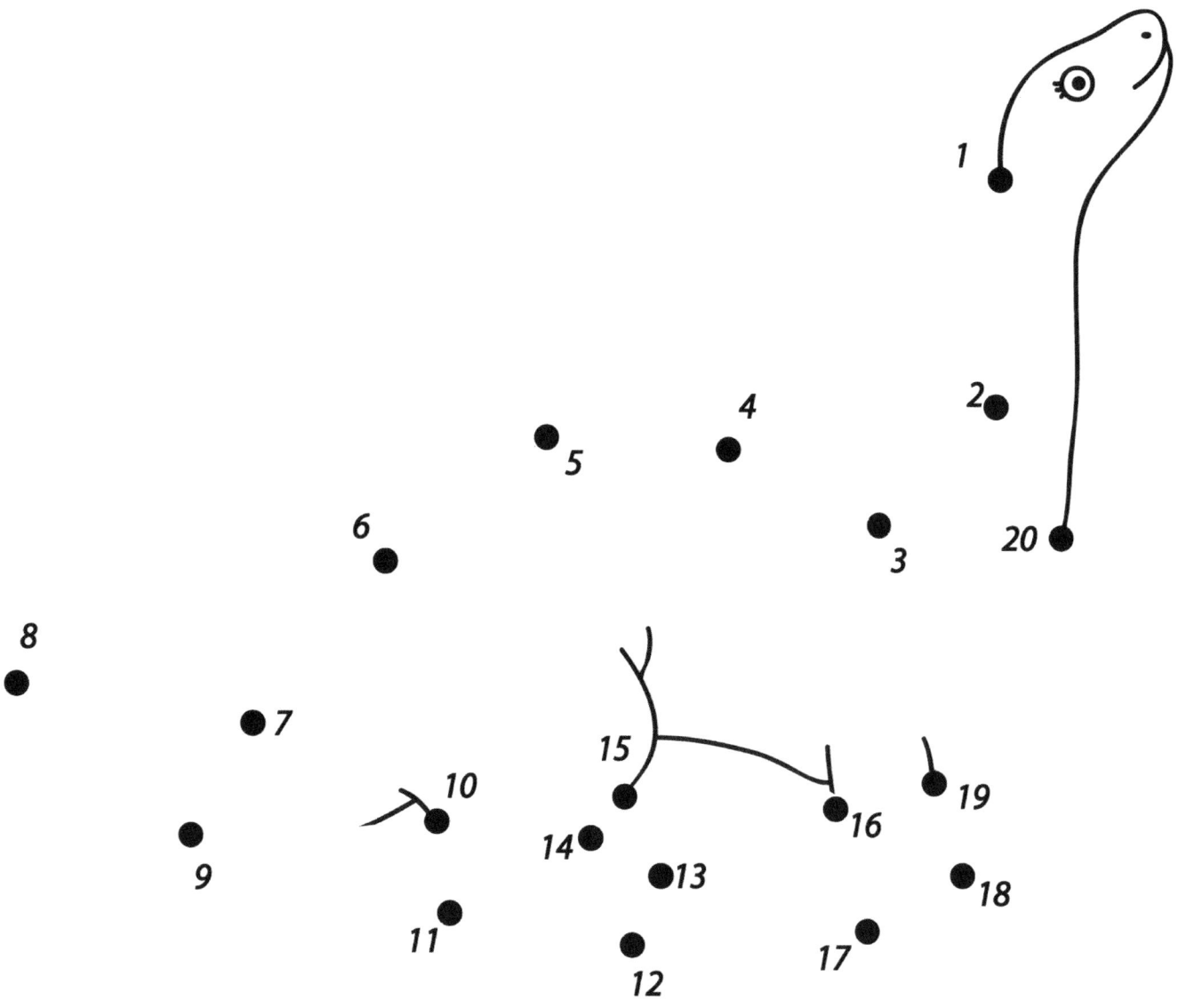

Printed in Great Britain
by Amazon